CAREER AS A

CONSTRUCTION MANAGER

SEVERAL OF TODAY'S MOST REWARDING AND LUCRATIVE CAREERS can be found in the construction management field. Construction managers (or CMs) plan, coordinate, budget, and supervise construction projects throughout the building cycle. They are sometimes called construction project managers (CPMs), general contractors, or simply project managers (PMs, a term also used to refer to non-construction project leadership, such as software development).

Construction managers may work for large corporations, residential and industrial building companies, government agencies, and nonprofit institutions. They may be direct employees of the company sponsoring the project; work for outside engineering, architectural, and construction contractors; or be self-employed. While the size of the individual project will vary substantially from a home renovation to an airport, for example, construction managers are found in virtually every industry, including aerospace, transportation and logistics, retail and homebuilding.

The median annual salary for construction managers averages almost $90,000, according to government statistics. Employment is expected to grow by at least five percent through the coming decade, due to a combination of new project demand and retirement of experienced managers. Construction professionals may work at corporate offices; at a single construction site; or commute between several job sites. Most managers work at least 40 hours per week. Some 40 percent of managers are self-employed.

Would a construction management career be right for you? Technical training and experience in the construction industry are required to get started. The profession also calls for good management skills, solid oral and written communication abilities, and analytical and planning skills. Are you a natural planner? Can you see the big picture – not just the immediate issues in front of you, but also the steps needed to reach a long-term goal? Are you good with science and math? Can you weigh competing alternatives to determine the best solution? Are you organized? If so, you may be well positioned to succeed as a construction manager.

While a four-year college degree is not mandatory, a growing number of employers look for a combination of formal training and on-the-job experience when hiring a new building CM. Additional certification from a professional organization can be important to advancing your career. Many construction managers are specialists in a certain industry (such as renovating schools or building new office parks). Managers must also keep up-to-date on new building materials and techniques, and use sophisticated software tools to track project costs and schedules.

If you have good analytical and interpersonal skills, you can enjoy a financially rewarding career in construction management. A combination of training, hard work, aptitude, and positive personal traits can help you achieve personal and professional satisfaction in the construction management field.

WHAT YOU CAN DO NOW

IF A CAREER IN CONSTRUCTION MANAGEMENT sounds interesting, there are steps you can take now to prepare yourself to enter the field – even while you are still in high school. Start with classes in mathematics and statistics. Computer science classes (such as Computer-Aided Design or CAD) can be helpful, both to learn the tools you will be using and to enhance analytical thinking. Good PC skills will be required on the job, as construction managers use such automated tools as project management software, email, spreadsheets and word processing software. Learning how to use tools like Microsoft Word, Excel and PowerPoint will be helpful, both for college courses and in your future career. Courses may be available at your school, through local colleges, or via online tutorials.

To learn more about the construction industry and the management profession, start reading industry publications such as *Construction Today*. Numerous magazines, blogs, and news sites are available online. Visit the websites of professional associations, such as the Construction Management Association of America (CMAA) and the Project Management Institute (PMI). Many associations also have a strong presence on social media, which may help you find local chapters and members in your area. If there is a local chapter geared towards construction managers, you may be able to visit a meeting or obtain a student membership. Networking with local professionals can help you learn about scholarships, internships, and entry-level job openings in your area.

Get a construction job, either part time or during summer school vacations. You may not be paid, and you may not do much of the actual building, being essentially a "helper," for the carpenters, masons and other professionals. You will learn from observing and talking with them. You will have a better chance of getting involved in smaller projects, like a home renovation in your neighborhood.

HISTORY OF THE CAREER

SINCE THE EARLIEST DAYS OF recorded civilization, human beings have designed and built buildings, roads, bridges, and homes. Many ancient construction projects required planning, execution and control while erecting monuments, some of which still stand today. The Great Pyramid of Giza (completed in 2570 BC) took 20 years and 10,000 workers to complete, while millions of people worked on the Great Wall of China (208 BC). The Giza project was particularly challenging, requiring close workforce management; detailed engineering plans to create the king's chambers inside the structure; and logistical difficulties mining and transporting granite 400 miles to the construction site. Master builders oversaw these early undertakings.

By the Middle Ages, European castles and cathedrals represented the pinnacle of construction innovation. While castles were originally designed as fortresses to repel invaders, they developed into grand royal palaces, while retaining a number of innovative defense measures designed to thwart intruders. Kings and noblemen would typically outline general plans for the castle, leaving the details to construction leaders who were architect-designers, materials purveyors, and workforce overseers.

The Industrial Revolution (1750-1850) created a new class of business managers; separation of the architect and construction management functions; and the development of current management theory. Early architects often managed the projects they designed, but eventually the separate profession of construction manager evolved. Sophisticated construction projects required experienced, capable engineers and managers who could focus on implementing the architect's vision. For centuries, building experience was enough to qualify an individual to lead a construction project. By the 19th Century, the master builder also needed to read and write, do arithmetic, be familiar with contemporary new building materials and techniques, and bring business and financial skills to the management position. For example, the Transcontinental Railroad (1896) that linked California to the East Coast was among the first government projects requiring advanced project management skills.

However, for the most part, architects, engineers, and master builders typically managed construction projects until the early 1900s. By that time, master builders were evolving into the new profession of construction manager – an expert in coordinating all the interlocking phases of a project – as project management itself became more sophisticated. Demands to complete complicated building projects on time and within budget would eventually drive the development of modern project management techniques that were required for complex endeavors such as computer software development and the US Space Program.

Henry Gantt, the "father of planning and control techniques," introduced the first widely-accepted project management tool in 1912: the Gantt chart (still popular with construction managers today). The Gantt chart was used during ship construction for World War I and in the 1931 construction of the Hoover Dam. The Associated General Contractors, the oldest and largest U.S. group of its type, was founded in 1918.

Early 20th Century construction management techniques are best reflected in the Empire State Building in New York City, which was the world's tallest building for several decades. General contracting firm Starrett Brothers and Eken managed some 3,500 workers from 60 different trades who worked seven million hours. The project was completed in April of 1931 after 13 months of construction, opening three months early and $18.3 million under budget.

By the 1950s, contemporary project management had taken root in the construction and engineering fields, introducing a targeted discipline with theories and tools that operated across multiple industries. In 1956, the American Association of Cost Engineers (now AACE International) was formed. The group brought together project management, cost estimating, cost management, planning, and scheduling experts.

The Project Management Institute (PMI) followed in 1969. PMI was created primarily to promote project management as its own distinctive discipline.

The Construction Management Association was founded in 1982, with the National Institute of Construction Management and Research following the next year.

The next revolution came with the computer age in the late 20th century. Computers were first applied to government and defense capabilities, spreading to private industries of all types. Computers brought new project management tools, computer-aided design (CAD) software to automate architectural drawings and blueprints, and risk management-forecasting programs that made construction management more efficient.

By the 2010s, cloud computing and mobile PM applications freed construction managers from their desktop computers, allowing them to oversee virtual project teams and work in far-flung locations. While these undertakings still require on-site management, mobile tools allow construction executives to take a broader view of budgets, costs, and resource management across multiple construction sites. As projects continue to grow larger, more technically complex, and span more countries, construction managers will be challenged to address a broad range of new opportunities for years to come.

WHERE YOU WILL WORK

CONSTRUCTION MANAGERS ARE FOUND working for companies of all sizes in cities and towns across the country. Construction firms that specialize in building schools, hospitals, shopping centers, apartment complexes, and residential subdivisions are leading employers of professional managers. These experts may work for private and publicly-traded companies, multinational oil and gas producers, government agencies, nonprofit entities, real estate developers, water and sewer utilities, military branches such as the US Army Corp of Engineers, and a wide variety of other enterprises. Self-employed construction managers may handle individual projects for those entities on a temporary contract basis, or run their own small companies that renovate historic buildings or build single-family homes, for example.

Most managers have a company office, particularly if a major corporation or a construction contractor employs them. However, the majority of construction managers also work from a field office at the construction site so they can closely monitor progress on the project and oversee daily activities. Those who manage multiple projects in different locations spend time traveling between sites in different cities, states, or even foreign countries.

Almost 20 percent of the more than 370,000 construction managers in the United States work for specialty trade contractors. Another 15 percent are employed in nonresidential building construction, while about 10 percent work in residential construction, and 10 percent are employed in heavy and civil engineering construction. About 40 percent are self-employed.

Managers spend long days (plus sometimes nights and weekends) on the jobsite or in the office. Most work full time, however, the need to meet deadlines and to respond to delays and emergencies often requires construction managers to work many hours. Managers may also be on call 24 hours a day.

California is the state with the largest number of construction managers with almost 30,000, followed by Texas (27,000), Florida (18,000), New York (14,000), and Ohio (9,000). Top metropolitan areas for construction managers include New York, Houston, Los Angeles, Chicago, Dallas, Washington, DC, Denver, Phoenix, and Boston.

Construction managers are not confined to large cities and populous states. They also work in small cities, farming communities, and rural areas, building churches, shopping malls, and highways that bring people together.

THE WORK YOU WILL DO

CONSTRUCTION MANAGERS (CMS) plan, coordinate, budget, and supervise building endeavors from start to finish. They strive to ensure that the project is completed on time, on budget, and to high quality standards. CMs coordinate many different types of construction projects, including highways, bridges, airports, factories, homes, hospitals, and government buildings. Regardless of the industry and the type of project, the CM is responsible to the owner of the project to deliver a successful project on time and within budget.

CMs may work on one construction site or oversee several sites simultaneously. On complex projects (such as an apartment complex with multiple buildings), there may be too much work for one person to effectively manage everything. In that case, a CM will engage, supervise and coordinate the work of one or more assistants. An alternative arrangement is for different assistant CMs to focus on particular phases of the project, with one top-level manager coordinating their efforts.

The main difference between a construction project manager and an assistant manager is the level of authority. The CM supervises one or more assistants. The assistant often carries out specific duties delegated by the CM (such as tracking project budgets, making sure overtime pay is properly calculated, or obtaining building permits), so the CM can focus on higher-level issues. CMs also have similar duties to a general construction superintendent. However, the CM typically has a broader range of management responsibilities, while the superintendent mostly oversees the foremen, who directly supervise individual workers. Sometimes CMs are called cost estimators, as they may spend more time on cost calculations than any other aspect of their job.

While each project is different, construction managers generally have the same basic duties during the construction process:

- Prepare budgets, estimate costs, and create timetables with target dates for accomplishing each set of work tasks.

- Create and submit contracts and bids for new projects.

- Specify project objectives, such as the scope of the work and the skills needed among team members.

- Hire, supervise and train employees on company procedures.

- Read blueprints and schematics.

- Maximize cost-effective and efficient use of labor, equipment, and materials.

- Use specialized computer software to estimate costs, order materials, lease equipment, handle delivery logistics, and assign work to various team members to comply with financial targets and scheduling goals.

- Chose the subcontractors who will do the work, and oversee their activities and those of their employees.

- Apply for the necessary construction permits and licenses.

- Work with planning commissions, building inspectors, fire inspectors, and other regulators to obtain project approval.

- Ensure the project complies with building codes, government regulations, and other legal considerations.

- Maintain compliance with safety guidelines and labor regulations.

- Coordinate the timely use of electricians, carpenters, bricklayers, plumbers, heating and air-conditioning installers, and similar trade workers.

- Provide the client with regular progress reports, including tracking actual costs against budgets and the work accomplished versus the schedule.

- Respond to emergencies, delays in construction progress, labor issues, material delays, and similar problems.

- Develop effective communications channels to resolve conflicts among team members, vendors, and suppliers.

- Provide information to supervisors, subordinates, and peers about progress and problems.

- Update the project owner and upper management with the current status of the construction work.

- Identify potential risks in the project and take steps to contain those risks.

- Schedule project team and owner/developer walk-throughs as needed.

When a project sponsor decides to undertake a construction project, architects and engineers start by developing the basic design, and then create detailed blueprints outlining building plans. CMs may have input to the design process while the project is still being sketched out and budgeted, on such topics as what materials to use and the type of land required. If the project involves renovating an existing building, the CM is more likely to get involved earlier in the process, since there will be many complex construction issues involved. If not part of the design team, the CM's duties start in pre-construction (when key personnel are chosen and site preparation begins) and procurement (hiring workers, ordering materials, and scheduling machinery). The construction phase – actually building the project – follows, continuing until the owner occupies the new facility.

CMs are primarily responsible for the overall building and implementation of a new structure. They must constantly balance costs, schedules, resources availability, safety considerations, compliance with building codes, and quality control measurements to ensure that the project meets its goals. While they have a considerable amount of authority, they may still consult with higher-ups (either their direct manager or an external client) to approve changes to construction plans that affect cost and scheduling targets. CMs may also collaborate with architects, engineers, building inspectors, and similar specialists when recommending significant alterations in the original plans.

Once the work begins, the CM's duties shift from planning towards monitoring and overseeing the process. The CM monitors progress on a periodic basis through meetings, status reports, time-tracking software, and similar tools and techniques. If the project is not advancing as scheduled, managers resolve roadblocks and update the project plan to reflect discrepancies between the expected versus results. They also identify and analyze requested changes in the original scope of the project – such as when the sponsor wants to cover concrete floors with hardwood – and manage those updates through a change control process.

Once the project is completed and ready for owner control, the final duties for a CM involve closing out the project. This may include paying outstanding invoices, closing bank accounts, providing employee performance feedback, conducting final audits, and turning the completed project over to the operational team who will run the completed school, hospital, shopping center, or other structure. Many projects are followed by a "lessons learned" review recapping what worked well and what could be done better in future projects.

The career path for becoming a construction manager varies widely among industries and organizations. An increasing number of companies favor holders of two- and four-year degrees over those with less education, as formal training exposes students to more options than can be gained on a few job sites. Still, while education is important, employers are particularly interested in a future CM with years of experience working on actual job sites. Those with experience plus training can often land supervisory CM roles. Internships, cooperative education programs, and previous work in the construction industry can provide the experience you will need.

Whether they began as carpenters, plumbers, ditch-diggers, or masonry apprentices, construction professionals at all levels of the company start out at the bottom and (sometimes literally) work their way up. Over time, as they gain experience and knowledge in multiple trades, they may be able to move up to positions of increasing responsibility, such as crew leader, team supervisor, assistant CM, and finally CM.

While titles and specific duties may vary among industries, there are similarities in CM duties and responsibilities among the basic types of construction projects. Each of these types follows similar CM disciplines, but they require specialized knowledge in specific industries. Those construction types are:

- **Agricultural,** which covers structures such as barns, silos, and animal sheds, plus special water supply requirements including wells and drainage.

- **Residential dwellings** such as houses, apartments, and townhouses.

- **Industrial buildings,** including steel mills, storage facilities, power plants, factories, oil refineries, and seaports.

- **Commercial structures** for private businesses, such as shopping malls, office buildings, hotels, theaters, resorts, and golf courses.

- **Institutional structures** for government and similar public organizations, including schools, hospitals, libraries, police stations, museums, and dormitories.

- **Heavy civil projects,** such as fortified military bases, dams, roads, bridges, airports, railroads, and transportation tunnels.

- **Environmental,** which had been a subset of civil projects, encompassing water and sewer plants, air pollution facilities, and solid waste management.

Despite the different focuses, most construction managers are able to move between companies and industries through their knowledge of a common set of duties, skills, tools, and computer programs. While having in-depth knowledge of airport construction, for example, would make some more attractive to a civil engineering contractor, this CM may also move to another sector with relative ease.

Shifting among different industries and types of construction is more common among managers who work on various projects with multiple clients, than for those employed by a corporation or specialty firm. These contract CMs may be self-employed, or they may work for a contracting company that arranges for them to provide professional services to a variety of clients. Some self-employed managers specialize in certain sectors, while others are generalists who can work in many fields.

THE PROFESSIONALS SPEAK

I Am a CM for a Large Contracting Firm

"I've been in this business a long time. I went into the military during the early 1960s and, thanks to my time working in our family construction business, I was assigned to the Army Corps of Engineers. My dad was a World War II veteran and, after his war ended, he came home to a nationwide construction boom – first projects for all the work that had been delayed during the war years, and then new interstate highways, schools and hospitals during the 1950s.

Fresh out of the service, I wanted to go back to the family business in South Carolina, where my dad was still getting work orders faster than he could handle them. However, he wanted me to go to college first and get a formal business education so we could find better ways to expand the company. I attended Clemson University on the G.I. Bill and earned a Bachelor of Science degree. I studied business management, accounting, construction science, marketing, and other topics we could use to run the family business more efficiently.

After college, I started out as an assistant CM to get more on-the-job training from the experienced CMs who were helping us keep up with the work flow. I learned a lot in college, but there's nothing to match the education you get from working side-by-side with experienced professionals in the field. It was invaluable, just watching them fit together all the people, machinery, materials and financing you need to plan a project – and then adjust the plans when the weather goes bad or all the carpenters get the flu. I served a true apprenticeship with those mentors, and I try to provide the same opportunities to young people who work for me today.

As I became a full CM in the 1960s, our industry faced a new set of challenges. Inflation skyrocketed, driving up the cost of borrowed money, labor, and everything else we needed. Construction prices soared, making competition among bidders tougher and adding more pressure for us to operate efficiently. All the contractors faced the same issues, so we had to figure out how to do things better yet cheaper than anyone else. Over time, we adjusted to a more demanding new environment. When the economy returned to 'normal' in the 1970s, we were better prepared to compete.

Since that time, we've weathered boom and bust cycles in the economy, new technology, tougher building codes, stricter safety standards, new labor laws, globalization of our materials suppliers, and too many other changes to mention. One of my brothers took over ownership of our company after our dad retired, and I'm now a senior vice president and lead CM. Even after all these years, I still get a thrill from turning over a finished office building or retail store to the new owner so they can move in and benefit from our work."

I Own My Own Green Construction Business

"I did not start out with plans for a construction management career. I had a yard mowing business at age 12, and then worked as a gopher on some construction jobs in my hometown during the summer while in high school and college. I learned a lot, assisting more experienced craftsmen, but I did not expect to spend my life on construction sites. My primary goal was to simply land a management position. I graduated from Providence College in Rhode Island in the early 1980s with a bachelor's degree in business management.

Soon afterwards, I went to work as a manager for a financial services company in Bridgeport, Connecticut. I spent three years learning about banking, insurance and securities, and managed a small team of office workers. However, I soon realized this was not the right industry for me. Sitting behind a desk in an office, it was difficult to feel like I was really making an impact on people's lives. I missed the immediate gratification of building a brick masonry wall or framing a house, and then watching a young family make it into their home.

One of our clients was a general contractor who came to our office for project financing and liability insurance. I told him once that I envied his career choice. 'Why don't you come work for me?' he asked. 'You seem to be a good manager, and you've worked at job sites before. I can show the rest of what you need to know.'

So that's how I ended up as a construction manager – an off-hand remark to the right person immediately after his assistant CM retired. I spent four years with the company in New Hampshire, learning the finer points of project planning and execution. There were no computers in the offices when I started. We sketched out buildings, drew blueprints, laid out project schedules, paid invoices, tracked hourly workers, and everything else, by hand. Today, we have automated tools to do all those tasks, so the efficiency is far beyond our old manual processes. But I'm glad I learned how to do everything 'the hard way' so I can make sure the computer-based tools are providing reasonable guidance. Technology is fantastic, but if you don't understand the process in your head, it can get you into trouble.

I quickly discovered that logistics is the heart of the construction business. We build apartment complexes in the Northeast. You have to coordinate a lot of moving parts – materials shipments, worker availability, equipment rentals, building inspections at key milestones, etc. All the components need to come together in the most efficient manner possible to save money, avoid conflicts, and finish by your deadline.

After four years, our company owner retired and sold his operation to a larger construction firm. I moved back to Rhode Island to go into business for myself. Energy-efficient housing was becoming popular at that point in the early 1990s, so I went into that niche of the residential market. I built environmentally-friendly homes and renovated existing houses to be more energy-efficient.

I've spent the last 25 years as a self-employed construction manager, building my own projects or subcontracting for larger firms to develop entire subdivisions. I've never regretted my career choice. Sometimes when the economy is down, our business suffers, but it always rebounds over time. I've learned to 'make hay while the sun shines' and save money to ride out the downturns. I'm proud of the company I built and of the homes we have built for the community."

PERSONAL QUALIFICATIONS

A CAREER IN CONSTRUCTION MANAGEMENT requires both technical knowledge (hard skills) and favorable personal attributes (soft skills). The basic technical skills for the construction industry can be learned in school and on the job. However, basic management skills and personal traits are also important to achieving a promising career as a CM.

Good communications skills are essential. You will do more than simply lead the team. You will be dealing with a variety of people on a daily basis – engineers, subcontractors, company owners, trade workers, vendors, architects, inspectors, and laborers. They all have different issues that need attention, and you will need to listen carefully to understand their concerns. Not everyone on the work site speaks fluent English, so clear speaking plus some knowledge of languages like Spanish, Russian, and Polish can come in handy.

To ensure common understanding, you must communicate understandably and effectively – both speaking and writing. There will be plenty of paperwork, including emails, status reports, performance reviews, and bids for new projects. The ability to gather, absorb, and share a vast amount of complex data effectively will help you function better at the job site.

Sound problem-solving and analytical skills are also important. You must understand the complex components that make up a building project and how they interact, in order to create a reliable schedule. That allows you to determine which tasks must be done first, what skills a person needs to perform that work, and a reasonable time frame for accomplishing each task.

You will need to be a critical thinker who relies on logic and deductive reasoning to find solutions for the challenges of planning and executing the construction project. Throughout any large undertaking, there will be challenges (such as sick employees, delayed materials delivery, and stormy weather) that require the manager to reassess the current situation and reshuffle priorities to keep things moving. Creative skills are required as you determine the most cost-effective way to finish the project on time with minimal disruption and conflict.

Ultimately, people make a project run, so people skills are also crucial for a successful construction manager. Every worker is a member of the team, and cooperation among those individuals helps keep the project on track. Good managers must build trust and appreciation within the workplace. You also need to be a good judge of others' abilities, as you often need to delegate tasks to trusted subordinates.   While you are the leader of the project, you are also a participant and a team member. This role requires you to be a good listener, a solid communicator, and an active participant in group settings.

Good organizational skills are also vital. You will need to juggle multiple projects with different deadlines, and use sophisticated project tracking software to make sure everything is proceeding as expected. You also need to manage your time and yourself in a professional manner, setting the example that you expect team members to follow.

Hopefully you will enjoy the construction process and the constantly changing industry. Many building projects use innovative technology, such as environmentally-friendly construction. You will spend a good deal of time during your career learning about new materials, new techniques, updated building codes, and innovative methods of managing projects.

ATTRACTIVE FEATURES

CONSTRUCTION MANAGEMENT IS A ROBUST FIELD offering a wide variety of opportunities to those entering the profession. The earnings, benefits, and personal potential along this career path are always moving up. New buildings continue to be erected and older ones are being renovated, which drives ongoing demand for skilled professionals to manage those projects. A vast range of options is available with a variety of employers, including large public multinational corporations, private companies, government agencies, and nonprofit institutions. Every growing industry spends money on construction, including such fields as energy, transportation, education, healthcare, and homebuilding.

If you like being in charge, solving problems, planning ahead, leading a diverse team, interacting with peers, and thinking creatively, you should find this career fulfilling. You will work with, and learn from, experts in many disciplines as you apply new building techniques and materials to construction challenges. Your work will have a positive impact on your firm, your clients, your community, and the people who live in your city and state. Success can bring personal fulfillment, professional growth, and the satisfaction that comes from a job well done. Few careers offer the opportunity to create something that will stand for decades!

Companies who need skilled, experienced construction leadership provide competitive salaries, attractive benefits, modern equipment, and advanced training to help attract the best candidates to their firms. Typical benefits include health and life insurance, paid vacation and sick days, retirement benefits, and company stock plans at larger firms. Many construction managers get a vehicle allowance or free use of a company-owned truck. These organizations also typically offer employees a well-defined path towards career advancement for those who wish to eventually manage multiple projects or become an executive.

Capable construction managers are highly regarded by colleagues throughout the organization and by the clients and owners who reap the benefits of their efforts. A manager who can deliver building projects on time, on schedule, and on budget is highly valued in corporate culture and the business marketplace. Successful management professionals have excellent prospects for moving into higher-level leadership roles.

You may decide to climb the corporate ladder, go into business for yourself as an independent construction manager, specialize in a certain niche (such as green buildings), or challenge yourself with increasingly larger and more complex industrial projects. Whichever path you choose, a career in construction management can be both financially and personally rewarding for many years to come.

UNATTRACTIVE ASPECTS

WHILE A CAREER IN CONSTRUCTION MANAGEMENT can be rewarding, it is also likely to be demanding and stressful. The downside of having broad authority over a project is the pressure and long hours accompanying that responsibility.

Your customers may be internal (fellow employees of your company) or external (your employer's clients). When a project falls behind or costs rise unexpectedly, they will expect you to control those variables quickly, efficiently, and accurately. Construction management deals with a number of interacting factors, including market costs, labor issues, materials shortages, safety mishaps, and bad weather. When things go wrong, getting the project back on track may not be as easy as it looks to an outsider. It may cost more money or take more time than is available. Those expectations can strain otherwise solid working relationships.

Construction managers are often called on to work long hours, nights and weekends. Many are on call 24/7 to respond to emergencies. Most managers are paid a set annual salary rather than an hourly wage, so there is seldom overtime pay. You also have to plan your personal life, your vacations, and other family time around an erratic work schedule.

Construction management requires juggling multiple complex factors.

Projects often last for years, and even basic home renovations can take several months. Such long endeavors can bring personal stress or even burnout, as it takes a long time to see the results of your work. You may not have the luxury of only working on one project at a time. Many managers oversee multiple projects that may not even be in the same geographic area. While you are in charge of the work site, you may still have to deal with demanding bosses, personality conflicts, unreasonable client expectations, and confusing government regulations.

Most managers work at the construction site itself. These places can be hot, noisy, dirty – and even dangerous, despite the best safety precautions. Some executive management positions may be based in a clean, safe, modern office environment, but they still often require visits to job sites. All managers are susceptible to long-term health issues arising from office work. For example, frequent computer use can also lead to eye strain, back pain, and repetitive motion injuries such as carpal tunnel syndrome.

Some construction managers must travel far from home to oversee projects. They may end up living for months in trailers at the construction site. Those managing multiple projects may commute daily between different locations. Long absences away from home and frequent travel may lead to family difficulties.

Managers also face constant demands for education – both to get started in the field and to maintain licenses or certifications. Some employers offer training courses, but others expect you to schedule (and possibly pay for) your continuing education needs.

EDUCATION AND TRAINING

WHILE YOU CAN BEGIN YOUR CAREER as a construction manager without a college degree or professional certifications, having higher education credentials on your résumé will enhance your ability to land that first supervisory position and advance in your career.

In today's employment market, it is helpful for construction managers to have a bachelor's degree in construction science, construction management, architecture, or engineering. More than 100 colleges and universities offer accredited bachelor's degree programs with majors in construction engineering, building sciences, construction science, and similar designations. Typical college curriculums cover project management and project control, construction methods and materials, design, cost estimation, contract administration, building codes and standards, statistics and mathematics.

Some of these universities also offer master's degree programs in construction management. Most construction management positions do not require a master's degree, particularly for entry-level positions. However, this level of training can help candidates advance from direct project management to executive positions with large contractors and multinational corporations.

More than 50 two-year colleges also offer construction technology and construction management courses. These can result in an associate degree, which combined with work experience is a desirable background for managers who supervise smaller projects. You can even get this training online or in evening classes.

Candidates with several years of work experience in the construction industry and a high school diploma may also move into construction management, especially working for themselves. On-the-job construction experience is even more important for those who do not have a college education. You can get this experience through internships, cooperative education programs, and previous work in the construction industry. Some construction managers become qualified solely through extensive construction experience, spending many years in carpentry, masonry, or other construction specialties before they move into management.

Practical on-the-job training offers several benefits. One is being able to learn new skills while working as an apprentice to more experienced crafts workers. You will be able to learn supervisory, technical, and organizational skills from experienced construction managers (plus architects, building inspectors, and engineers) in a real-world environment. You can try different trades – plumbing, wiring, pouring concrete, or erecting steel frames. You may also need to obtain trade licenses for certain jobs, such as plumber or electrician.

In addition to at-work training, you can pursue self-learning through books and classes to gain additional industry and management knowledge. Both on-the-job and independent learning will better prepare you for college or advancing to a higher position in construction.

Which college should you attend for your undergraduate and graduate work? First, look for schools that offer industry-accredited programs. The Construction Management Association of America (CMAA) accredits college-level management curriculums, while the Accreditation Board for Engineering and Technology (ABET) focuses on those fields.

Several national surveys rank the best schools for construction management and engineering training. For example, the rankings by US News and World Report list these top civil engineering undergraduate schools:

- University of Illinois

- Georgia Institute of Technology

- University of California-Berkeley

- University of Texas

- Massachusetts Institute of Technology (MIT)

- Purdue University

- Stanford University

- University of Michigan

- Virginia Tech

- Cornell University

A Study.com ranking of top construction management schools lists Purdue, plus Drexel University, University of Denver, University of Florida, University of Kansas, and Western Illinois University. Other surveys include Clemson University and the University of Washington as top construction management schools.

Among graduate schools, the US News and World Report ranking also includes California-Berkley, Illinois, Texas, Georgia Tech, Stanford, Purdue, MIT, Michigan, Virginia Tech, and Cornell as the top universities.

After college, there are a number of professional certifications available to advance your career. Some states require construction managers to be licensed. However, industry certifications are typically voluntary. Certain employers may require specific certifications in some situations, but for the most part, they primarily serve as evidence to future employers that you are qualified to be a construction manager.

Several types of certification are available from various organizations. They typically require a combination of on-the-job experience, self-study, and passing one or more examinations. The American Institute of Constructors (AIC) offers both the Associate Constructor (AC) certificate and the more advanced Certified Professional Constructor (CPC) designation. The Certified Construction Manager (CCM) designation awarded by the Construction Management Association of America (CMAA) includes testing on risk management, legal considerations, and professional duties for a manager.

Even after you receive certification, you will be expected to pursue periodic continuing education to maintain and broaden your skills. This training could take place in a formal setting (such as company-sponsored classes), or through self-study using books or online classes. The continual introduction of new construction materials and practices, plus new management techniques, mean there will always be something new to learn as a construction manager.

EARNINGS

THE MEAN ANNUAL EARNINGS OF CONSTRUCTION MANAGERS are $95,000. Construction managers rank somewhat lower than the $98,000 median wage for all management occupations. Among construction managers, the lowest paid 10 percent receive about $55,000, while the top 90 percentile average more than $155,000 in annual salary.

The top-paying industry for construction managers is heavy and civil engineering with median annual wages of $90,000. Nonresidential building construction is second at $85,000, followed by specialty trade contractors ($80,000) and residential building ($78,000).

The state of New Jersey offers the highest pay for construction managers with an average mean salary of about $140,000. Alaska is second at $125,000, followed by Rhode Island ($120,000), Pennsylvania ($115,000), and New York ($110,000).

Government statistics indicate that most construction managers work full time. There can be deadlines, delays and emergencies that require construction managers to work many hours. Many managers are also on call 24 hours each working day. Salaried construction managers may also receive overtime pay.

About 40 percent of construction managers are self-employed, making their earnings dependent on the amount of business they generate.

Bear in mind that salary is only one factor while considering the total compensation package for a position. Other benefits generally include medical, dental and life insurance; performance bonuses; paid holidays and vacations; retirement plan contributions; and vehicle allowances or use of a company truck.

OPPORTUNITIES

EMPLOYMENT OF CONSTRUCTION MANAGERS is expected to increase by about five percent to over 390,000 within the coming decade. That growth rate is about the same as the six percent rise expected in all management occupations.

There will continue to be a need for managers as construction activity expands. Population and business growth will result in the construction of new residences, office buildings, retail outlets, hospitals, schools, restaurants, and other structures over the coming decade.

The trend towards green building – constructing new structures or retrofitting existing buildings to make them more energy efficient and eco-friendly – should also create more jobs.

The need to ensure that projects are completed on time and within budget will also drive brighter job prospects for managers. Also, as technology and the construction industry become more complex, the need for managers with specialized expertise to provide oversight and leadership increases. The Department of Labor notes that job opportunities will be particularly good for those with construction experience plus a bachelor's degree in construction science, construction management, or civil engineering.

Demand for new jobs will continue to rise for the foreseeable future. However, the pending retirement of substantial numbers of construction managers over the next 10 years will also create more job openings to replace retirees. As a result, the number of jobs available is likely to be greater than the number of qualified applicants through the mid-2020s.

The construction industry is particularly sensitive to economic trends. There may be periods of unemployment when the overall level of construction falls. These will be followed by peak periods of building activity resulting in abundant job opportunities for construction managers.

Money Magazine ranks construction project managers and construction estimators among its Top 10 Jobs in America. The report notes, "Thanks to a commercial building boom for healthcare facilities, big box stores, malls, and high rises, demand by large commercial construction companies has outstripped the ready supply of estimators and project managers." Those with specialized experience (such as hospital construction) "can command a premium in pay, since they can hit the ground running on similar projects," the magazine added.

The states with the highest concentration of construction manager jobs are Alaska, Hawaii, Colorado, Oregon, and Nevada. The metropolitan areas with the most positions are New York, Houston, Los Angeles, Chicago, Dallas, Washington, Denver, Phoenix, Boston, and Anaheim.

GETTING STARTED

ARE YOU READY TO PURSUE A CAREER as a construction manager? Deciding on your future path is a good starting point. However, there are several other steps you can take over the coming weeks and months to start preparing for your professional career.

Begin by gathering more detailed information about the specific role you want to pursue in construction management. For example, would you like to manage large projects like highways and shopping centers, or build single homes for families? Determine how you can begin working towards that goal and how much training you will need to get started. Written information about construction management can be found at school and public libraries, through professional associations, from websites for colleges and universities, and via high school guidance counselors. The Internet puts a vast range of data at your fingertips from government agencies, private companies, and other sources. Trade magazines offer current information. The business sections of most local newspapers report which skills are in demand by local employers, which companies are actively building, and which industries anticipate significant growth in your state.

Once you determine the education you will need for a particular job, look for colleges and universities that can provide that training. Also consider how you could get some experience to learn more about that sector of the construction industry. Any entry-level summer job where you help build an office or renovate a home will introduce you to the ups and downs of daily life in the industry. Even a part-time job upgrades your résumé while you earn a paycheck. Each construction-related position demonstrates industry knowledge and aptitude to a prospective employer.

Spend some time talking to people who already work in the construction management field, or with those who work closely with the profession (such as architects, general contractors, or building inspectors). Ask these experts what skills, training, and experience are most valuable for applicants seeking an entry-level position. You can find these individuals through professional associations, at job fairs, or by contacting major builders. Industry associations have plenty of information about the profession as well as local internships, apprenticeship programs, and job openings. Many of these groups have outreach programs that specifically target students who are interested in the industry. Some have student memberships that allow you to attend sessions at a reduced cost.

In addition to professional resources, call on your personal network for support and advice. Discuss your plans with family and friends, and seek feedback on whether construction management would be an appropriate choice for your personal strengths and interests. Do not forget to include your school counselor in the decision-making process. Counselors can share helpful information about scholarships, employment, and networking opportunities.

Once you gather your data, it is time to give careful thought to whether a career in construction management feels right for you. Are you comfortable with the education requirements – higher education, on-the-job training and possibly ongoing certification? Can you visualize a project from beginning to end, and the steps required to reach a goal? Are you a team player who likes to solve problems? Do you communicate well with fellow students and teachers, in both the spoken and written word? Would you be comfortable managing teams of men and women who may be older and more experienced than you?

The most important determination is whether you can see yourself happily pursuing a successful career in construction management. If so, start taking those first steps today towards a rewarding, fulfilling career!

ASSOCIATIONS

■ **Accreditation Board for Engineering and Technology**
www.abet.org

■ **American Council for Construction Education**
www.acce-hq.org

■ **American Institute of Contractors**
www.professionalcontractor.org

■ **American Society of Civil Engineers**
www.asce.org

■ **Associated Builders and Contractors**
www.abc.org

■ **Associated General Contractors of America**
www.agc.org

■ **Associated Schools of Construction**
www.ascweb.org

■ **Construction Financial Management Association**
www.cfma.org

■ **Construction History Society of America**
www.constructionhistorysociety.org

■ **Construction Industry Institute**
www.construction-institute.org

■ **Construction Management Association of America**
www.cmaanet.org

■ **Design-Build Institute of America**
www.dbia.org

■ **National Institute of Construction Management and Research**
www.nicmar.ac.in

■ **Project Management Institute**
www.pmi.org

■ **US Green Building Council**
www.usgbc.org

PERIODICALS

■ **Builder**
www.builderonline.com

■ **Construction Business Owner**
www.constructionbusinessowner.com

■ **Construction Dive**
www.constructiondive.com

■ **Construction Executive**
www.constructionexec.com

■ **Construction Journal**
www.rics.com

■ **Construction Management**
www.constructionmanagement.com

■ **Construction Today**
www.construction-today.com

■ **Construction Week Online**
www.constructionweekonline.com

■ **Constructor Magazine**
www.constructormagazine.com

■ **Engineering News Record**
www.enr.com

■ **Equipment World**
www.equipmentworld.com

■ **Journal of Construction Engineering and Management**
www.ascelibrary.org

■ **Project Management Journal**
www.pmi.org/Knowledge-Center

■ **Sustainable Construction**
www.forconstructionpros/magazine/susc.com

WEBSITES

■ **Captera Construction Management Blog**
www.blog.captera.com

■ **CareerBuilder**
www.careerbuilder.com

■ **Careers in Construction**
www.careersinconstruction.ca

■ **Constructonomics**
www.constructonomics.com

■ **Contractor Talk**
www.contractortalk.com

■ **Dice**
www.dice.com

■ **Hard Hat Chat**
www.commericalconstructionblog.com

■ **National Center for Construction Education and Research**
www.nccer.org

■ **Project Management Salary Survey**
www.pmi.org

■ **Who's Who in Building and Construction**
www.thewhoswho.build

CAREERS INTERNET DATABASE
www.careers-internet.org

www.ingramcontent.com/pod-product-compliance
Lightning Source LLC
Chambersburg PA
CBHW061239180526
45170CB00003B/1370